Abel Hernández-Muñoz

DEMOGRAPHICS OF THE SIERRA ESPIRITUANA PALM IN BAMBURANAO, CUBA

Abel Hernández-Muñoz

DEMOGRAPHICS OF THE SIERRA ESPIRITUANA PALM IN BAMBURANAO, CUBA

Population study of the palm Gaussia spirituana Moya et Leiva (1991), in the mountainous massif of Bamburanao, Cuba.

ScienciaScripts

Imprint

Cover image: www.ingimage.com

This book is a translation from the original published under ISBN 978-620-2-16288-3.

Publisher:
Sciencia Scripts
is a trademark of
Dodo Books Indian Ocean Ltd. and OmniScriptum S.R.L publishing group

120 High Road, East Finchley, London, N2 9ED, United Kingdom
Str. Armeneasca 28/1, office 1, Chisinau MD-2012, Republic of Moldova, Europe
Printed at: see last page
ISBN: 978-620-7-85111-9

INDEX

INTRODUCTION

Palms are identified with tropical environments, be they rainforests or broad savannahs, the high mountains of the tropics or oases.

They are plants of great economic importance in the world, and millions of people in tropical and subtropical areas of the planet depend on them, directly or indirectly, for their survival.

There are many species of cultivated palms planted outside tropical countries. Mostly in the Mediterranean climate zones, where a large variety of them can live outdoors. In temperate and boreal countries, on the other hand, they are usually grown in greenhouses.

This group of trees is constantly increasing due to the continuous discovery of new species and to the modifications that the classification undergoes as this family is studied in more depth. At present, the number of genera comprises about 180, grouping about 2,400 species. We have taken as a reference the classification published in Genera Palmarum. The evolution and classification of palms (2008), written by John Dransfield, Natalie W. Uhl, Conny B.Amussen, William J. Baker, Madeline M. Harley, and Carl E. Lewis.

The family Arecaceae (Arecaceae) is the only family of the order Arecales (former order Princes) of the Monocotyledons and includes all the palms; in fact, it was called Palmaceae before acquiring its present name. It is a family of great originality and represents a quite natural and homogeneous group, since the various species combine arboreal bearing, wide leaf blades and a very well developed vascular system in all its vegetative organs, formed by numerous and small separate vascular bundles, each wrapped by a fibrous sheath. In their stems or stipes are frequent scars left by the leaves when they fall. They are usually plants with unbranched stems (monocaules), although some, such as Hyphaene and Nypha, may have dichotomous branching. Also, the stipes are usually topped by a characteristic crown of leaves, each with a broad clasping base. The leaves are quite complex and can be of various shapes and sizes. The inflorescences are also very diverse in shape. The flowers are unisexual or bisexual and the plants are monoecious or dioecious, although in some cases they can be polygamous. Pollination can be anemophilous (by wind), although most commonly it is entomophilous (by insects); some species are pollinated only by coleoptera (coleopterophilous). The fruit is fleshy, a monosperm berry or drupe.

This botanical family includes the genus Gaussia, which has two species in Cuba: *Gaussia princeps* Wentland (1865) from the mogotes of Viñales, province of Pinar del Río; and *Gaussia spirituana* Moya et Leiva (1991), distributed in the mountainous massif of Bamburanao, in the center-north of Cuba.

The pinareña species is better known, but there are no data on the demography of *G. spirituana* Moya et Leiva (1991), an endemic palm threatened with extinction, and only a few a priori population estimates are known; therefore, the general objective of this research is to provide information on the main demographic characteristics of this species.

The general objectives were:

1. Specify the localities where this palm is found,

2. To quantify the population numbers,

3. Determine the age groups of the demo,

4. Determine the type of distribution of the same.

5. Perform a vegetation profile where the plant lives.

BIBLIOGRAPHIC REVIEW

Arecaceae

Arecaceae

The coconut palm, *Cocos nucifera*

Taxonomy **Kingdom**: Plantae

Division: Angiospermae

Class: Monocotyledonae

Subclass: Commelinidae

Order: Arecales

Arecaceae Schultz Sch.

Family: **Palmae** nom.cons.) family no. 76 in LAPG III 2009[1

Subfamilies

- Arecoideae
- Calamoideae
- Ceroxyloideae
- Coryphoideae
- Nypoideae
- Phytelephantoideae

Arecaceae (scientific name **Arecaceae**, synonym **Palmae**), are a family of monocotyledonous plants, the only family of the order **Arecales** (synonym **Principales**). They are usually referred to as **palms** or **palms**. Individuals of this important family are easy to recognize visually, although there may be confusion with species of the families **Cycadaceae** and **Zamiaceae** due to morphological similarities. They are woody plants (but without secondary trunk growth, only primary). Despite being monocotyledonous many of them are arborescent, with large leaves in crown at the end of the stem, usually pinnate (pinnatisect) or palmate (palmatisect). Its flowers have 3 sepals and 3 petals, and are arranged in inflorescences with one or more spathes. The fruit is fleshy: a berry or a drupe. They are widely distributed in tropical to temperate regions, but mainly in warm regions.

The family was recognized by modern classification systems such as the APG III classification system (2009)[2] and APWeb (2001 onwards).[3] Traditionally it was also recognized in other classification systems due to its distinctive morphological characters. In these classification systems, they are placed in their own monotypic order Arecales, in the subclass Commelinidae.

Palm trees include species of economic importance and species of ornamental value, as well as others such as the coconut palm, oil palm, date palm, palm heart, rattan, carnauba wax, raffia, among others.

In the world they grow as species typical of tropical zones, there are concentrations of them in countries such as Madagascar. Colombia is the country with the largest number of varieties and one of them is the national tree. Additionally, there are several botanical gardens specialized in palms and they are often called palmetum. Some of these collections include the Palmetum de Santa Cruz de Tenerife and the Palmeral de Elche in Spain, the El Palmar National Park in Entre Rios, Argentina, and to a lesser extent, the Caracas Botanical Garden in Venezuela, and the National Botanical Garden of Cuba in Havana, Cuba. There are also important collections of palms in the Molino de Inca Botanical Garden in Torremolinos, (Malaga) and the La Concepcion Botanical Garden in Malaga.

Description

***Copernicia sp.* palm grove in Romero, Manatí, La Sierpe, south of Sancti Spíritus, Cuba.**

Trees or shrubs with unbranched trunks or rarely,[4] occasionally long rhizomatous herbs, or non-branching climbing palms (e.g. *Calamus*). Plant sex is variable. Secondary growth is absent. Stem apex with a large apical meristem, leaves develop helically. Tannins and polyphenols often present. Varied hairs, and plants sometimes spiny due to modified leaf segments, exposed fibers, pointed roots, or petiole growths.[5]

The root system is always adventitious with origin in the hypocotyl and lower nodes of the stem, since the primary seminal root is replaced at an early stage of seedling development, so the root of the palm is fasciculate (fibrous), with abundant ramifications, generally short and very dense, forming a bulb at the base of the trunk, which provides mechanical support in addition to the functions of absorption of water and mineral elements. Palms may present mycorrhizal symbiosis.

The stem is usually arborescent with a single unbranched trunk (dichotomously branched in *Hyphaene*) or in a cespitose clump of erect stems, or in a dichotomously branched erect rhizome (*Nypa*), or in a slender elongated bamboo-like but supportive stem (rattans). In some species it reaches more than 30 m in height.

Typical leaves are rather large, alternate and whorled (rarely distichous or tristichous), often clustered in a terminal crown (acrocaulis), but sometimes well separated, entire, sheathing at the base, with an elongate, erect petiole (sometimes referred to as a pseudopetiole) between the sheathing base and the lamina. In arborescent taxa the sheathing bases of adjacent leaves may overlap each other, forming a capital at the apex of the trunk. Leaves may be simple, usually divided into pinnately or palmate as the leaf expands, and at maturity appearing palmately lobed (with segments radiating from a single point), costapalmately lobed (with more or less palmate segments diverging from a short central axis, or "costa"), pinnately lobed or compound (with a well-developed central axis bearing pinnate segments), or rarely twice pinnately compound. Sometimes bifid. With leaflets converted into spines present in some taxa. Lamina "plicate", and segments either induplicate (V-shaped in cross section), or reduplicate (Λ-shaped in cross section), each segment with more or less parallel to divergent veins. The petiole, in taxa with palmate leaves, with a flap ("hastula"), between petiole and lamina. With soft tissues often decaying to reveal varied fiber patterns. Without stipules. Leaves typically ligulate, with the appendage, the ligule, at the inner junction between the lamina and petiole. Venation is pinnate.
-or parallel-clapping.

Palm trees in the Botanical Garden of Sancti Spíritus, Cuba.

Determinate or indeterminate inflorescences, panicles or spikes of solitary flowers or cymose units, typically axillary or also terminal, with small to large persistent and deciduous bracts. The inflorescences emerge from below (infrafoliar) or between (interfoliar) or above (suprafoliar) leaves. The peduncle has below a profilium often large with 1 to numerous spathes.

Flowers bisexual or unisexual (and then monoecious to dioecious plants), radial, usually sessile, with perianth usually differentiated into calyx and corolla, hypogynous.

Usually 3 sepals, separate to connate, usually imbricate.

Usually 3 petals usually, separate to connate, imbricate to valvate.

Stamens 3 or 6 to numerous, filaments separate to connate, free or adnate to petals. Staminodes present in some species. Anthers longitudinal, rarely poricidal in dehiscence.

Pollen usually monosulcate.

Carpels usually 3, but occasionally as many as 10, sometimes appearing as one, separate to connate. Ovary succumbent, usually with axillary placentation, but placentation variable. Styles, if present, separate or connate, stigmas sessile or at tip of styles, stigma varied. Ovules 1 per locule, anatropic to orthotropic, bitégmic.

Nectaries in the septa of the ovary or without nectaries.

The fruit is a drupe, usually single-seeded, often fibrous, or rarely a berry. Rarely dehiscent. Some have external scales (Calamoideae), hairs, stingers, or other protective structures.

Seeds usually one per fruit and with endosperm rich in oils or carbohydrates (hemicelluloses), sometimes ruminated. Starch absent.

Ecology

Entrance gate to the Botanical Garden of Sancti Spíritus, a jata palm on the right.

Widely distributed in tropical and subtropical regions, mainly in places with high humidity, with more than 2400 mm of average annual rainfall, more than 160 days with rain and more than 21 °C. Because of their abundance, they are often ecologically important where they are present. They are also found in temperate zones (e.g. *Chamaerops*), survive in desert environments (*Phoenix spp.)*, from tropical forests to mangroves (*Nypa fruticans*), and from sea level (*Cocos nucifera*) to very high altitudes (*Trachycarpus*).[5][6]

Worldwide there are more than 2,400 species, belonging to 27 tribes in five subfamilies. About 790 species grow wild in the Neotropics (Dransfield et al. 2008). The most species-rich Neotropical regions are found in the Chocó region, where up to 83 species can be found in a grid of about 10,000 km², followed by the Mesoamerican Isthmus region (Panama and Costa Rica) (Bjorholm et al. 2005). Colombia alone has 289 species grouped in 66 genera, which makes it the richest country in the Americas in terms of palm species, and it also has the highest number of endemic species with a total of 33 species, equivalent to 15% of the total number of palms in the country. Unfortunately, Colombia also has the highest number of threatened palms in the Americas with 30 species in some category of danger (17 of which are endemic). The most representative genera in Colombia are: *Astrocaryum, Bactris, Chamaedorea, Desmoncus, Euterpe, Geonoma, Mauritia, Oenocarpus* and *Syagrus.*[7]

Palm flowers are usually pollinated by insects, especially beetles, bees and flies. Nectar is often used as a reward for pollination (Henderson 1986).[8]

Palm fruits are usually fleshy and dispersed by a wide variety of mammals and birds, although some (such as *Nypha* and *Cocos*) are dispersed by water and float on ocean currents (Zona and Henderson 1989).[9]

Phylogeny

A large body of work over a period of about 30 years has clarified our understanding of palms (see for example Dransfield 1986,[10] Dransfield and Uhl 1998,[11] Henderson 1995,[12] Henderson *et al.* 1995,[13] Moore 1973,[14] Moore and Uhl 1982,[15] Tomlinson 1990,[16] Uhl and Dransfield 1987,[17] Zone 1997[18]). Dransfield *et al.* (2005[19]) presents a clarification of the family based on molecular relationships (see especially Asmussen *et al.* 2006).[20]

Arecaceae is easy to recognize and monophyletic. Palms are easily identified although there would be no consistent ((synapomorphies)) for the family. Uhl and Dransfield (1987) and Uhl *et al.* (1995[21]) have identified two main diagnostic characters: 1) "woody" stems (due to the presence of fibrous sclerenchyma, not secondary growth), and 2) pleated leaves on buds and subsequent division in most groups.

Calamoideae has pinnate to palmate leaves and distinctive fruits that are covered with imbricate reflexed scales (a synapomorphic character). Notable genera are *Raphia*, *Mauritia*, *Lepidocaryum*, *Metroxylon*, and *Calamus*.

Nypa (Nypoideae) has a prostrate stem that is dichotomously divided, and erect, pinnate leaves and undifferentiated tepals. Fossils are known from Europe and early Tertiary America.

Phylogenetic analyses of multiple DNA sequences show that the subfamily Calamoideae is sister to all other palms. *Nypa* (the only genus of Nypoideae), a distinctive genus of Asian and western Pacific mangrove communities, would remain sister to all other palms (except Calamoideae). Then *Nypa* and Calamoideae form a paraphyletic complex, with usually pinnate and reduplicate leaves, while the rest of the genera, with usually costapalmate or palmate and induplicate leaves - the Coryphoideae - form a monophyletic group (Hahn 2002,[22] Uhl *et al.* 1995).[21]

Arecoideae has pinnate leaves and flowers in groups of 3 (triads), with a carpellate flower surrounded by two staminate flowers (probably a synapomorphy, but lost in some subgroups). Within Arecoideae, a few well-defined monophyletic groups are evident.

Hyophorbeae (which has for example *Chamaedorea*, *Hyophorbe*), has imperfect flowers in lines.

Cocoeae have inflorescence associated with a persistent, large, woody bract and fruits with stone-like, triporate endocarp, and include genera such as *Elaeis, Cocos, Syagrus, Attalea, Bactris, Desmoncus,* and *Jubaea.*

Iriarteae (which has for example *Iriartea, Socratea*) has "stilt" roots, and leaf segments with blunt apices and divergent veins.

Most Arecoideae are placed within a heterogeneous Areceae (Baker *et al.* 2006[23]), representative genera include *Areca*, *Dypsis*, *Wodyetia*, *Veitchia*, *Ptychosperma*, and *Dictyosperma*. These palms sometimes have a structure consisting of a series of large or overlapping leaf bases, which resembles a vertical extension of the stem.

Coryphoideae are traditionally divided into the first 3 tribes below, a fourth is included here:

- The monogeneric Phoeniceae (*Phoenix*, date palms) have distinctive pinnate leaves in which the basal segments are like thorns.
- Borasseae (containing for example *Latania, Borassus, Lodoicea,* and *Hyphaene*) have staminate flowers embedded in thickened inflorescence axes.
- Most of the genera of Coryphoideae are placed in a third tribe, the heterogeneous "Corypheae". Notable genera include *Chamaerops, Rhapis, Licuala, Copernicia, Corypha, Washingtonia, Serenoa, Livistona, Rhapidophyllum,* and *Acoelorraphe*, and are difficult to characterize as a whole. Genera such as *Sabal, Thrinax*, and *Coccothrinax* are phenotypically similar, and have been included here, but their inclusion would make the tribe non-monophyletic.
- Caryoteae (fishtail palms), including for example *Caryota* and *Arenga*, form a distinctive clade within Coryphoideae (Asmussen *et al.* 2000,[24] Asmussen and Chase 2001,[25] Hahn 2002[22]) because of their flower triads (which evolved in parallel with those of Arecoideae). This group has induplicate, bluntly blunt leaf segments with divergent veins.

Taxonomy

The family was recognized by APG III (2009[2]), Linear APG III (2009[1]) assigned the family number 76. The family had already been recognized by APG II (2003).[26]

In the APG III system it is in the monotypic order Arecales, subclass Commelinidae, class Monocotyledoneae. The APG II classification system, which does not employ formal names above the order level, placed it within the Commelinidae clade, which was maintained from the 1998 publication of the APG system to the present.

Previous systems placed the group in the subclass Arecidae (Cronquist, 1981) or in the superorder Arecanae (Dahlgreen, Thorne).

The name Arecales, formed according to the rules of the International Code of Botanical Nomenclature from the genus *Areca* type (which includes the betel palm, *Areca catechu*) is of relatively recent use. The traditional nomenclatures used the descriptive name Principales, from Latin *the first ones.*

As an exception to the priority rule, the old names used by Linnaeus can still be used for this family: Palmae for the family and Principales for the order. The names Palmáceas or Palmaceae are rejected by the International Code of Botanical Nomenclature. They are called palms in Spain, Uruguay, Argentina and Chile, and palms in most countries of America.

The extensive family has 200 genera, 2780 species. The most represented genera are *Calamus (*370 species), *Bactris (*200 species), *Daemonorops (*115 species), *Licuala (*100 species), and *Chamaedorea (*100 species).

The following are some genera of the family, with some of their species, their botanical authorship and their common names, as used in different Spanish-speaking countries.

- *Aiphanes*
  - *Aiphanes caryotifolia* (H.B.K.) J.C.Wendl. - Mararay (Venezuela, Ecuador and Colombia)
- *Areca*
- *Arenga*
- *Astrocaryum*
  - *Astrocaryum aculeatum* G.Mey. - cumare (Venezuela and Colombia)
  - *Astrocaryumstandleyanum* -Guerregue , Wérregue, Chunga (Chocó).
- *Attalea*

- *Bactris*
    - *Bactris gasipaes* Kunth - chontaduro (Peru, Colombia and Venezuela)
    - *Bactris brongniartii* - Cubarro, Cubarro Palm, Apricot.
    - *Bactris guineensis* (L.) H.E.Moore
- *Borassus*
- *Ceroxylon*
- *Cocos nucifera* L. - coconut tree
- *Copernicia*
    - *Copernicia tectorum-Sará*, Palmiche (Colombia and Venezuela)
- *Chamaedorea*
    - *Chamaedorea pauciflora* -Iakake (Miraña)
- *Chamaerops*
    - *Chamaerops humilis* L.: Palmito, Margalló (Spain, in Spanish and Catalan)
- *Desmoncus*
    - *Desmoncus polyacanthos* -Matamba, Bejuco alcalde (Venezuela and Colombia)
    - *Desmoncus* mitis-Atajadanta (Amazonas), Bejuco mayor
- *Elaeis*
    - *Elaeis guineensis* Jacq. - Oil palm, Oil palm, African Palm
    - *Elaeis oleifera* (H.B.K.) Cortes - Nolí (Venezuela and Colombia)
- *Euterpe*
    - *Euterpe oleracea* - Murrapo, Palmito, Naidí, Palmiche, Manaca Colombia and Venezuela)
    - *Euterpe precatoria* - Asaí, Palmiche, Manaca Palm (Colombia and Venezuela)
- *Geonoma*
    - *Geonoma deversa* - Goguire de centro de monte (Uitoto), Tataba (Miraña)
    - *Geonoma macrostachys* - Ucsha (Ecuador), Palmiche (Perú)
- *Iriartea*
    - *Iriartea deltoidea* - Bombona, Bombonaje, Corneto, Trompeto, Barrigona, Cachudo (Venezuela and Colombia)
- *Jessenia* H.Karst.
    - *Jessenia bataua* Burret Correct name: *Oenocarpus bataua*
- *Jubaea*

- *Jubaea chilensis* (Molina) Baill. (Chile)
- *Jubaeopsis*
- *Jubaeopsis caffra* Becc.
- *Howea*
- *Howea forsterana* (C.Moore & F.Muell.) Becc.
- *Leopoldinia*
- *Leopoldinia* piassaba- Chiquichiqui (Colombia), Málama (Vaupés), Piassava (Brazil), Fibra
- *Lepidocaryum*
- *Lepidocaryum tenue* - Pui, Caraná, Erere (Uitoto)
- *Livistona*
- *Livistona australis* (R.Br.) Mart.
- *Lodoicea*
- *Lodoicea maldivica* Coco de mer (Seychelles), Coco de mer (Seychelles), Coco de mer (Seychelles), Coco de mer (Seychelles).
- *Manicaria*
- *Manicaria saccifera* Gaertn. - Napa (Panama and Colombia) or Cabecinegro (Colombia)
- *Mauritia*
- *Mauritia minor* Burret
- *Mauritiaflexuosa* L.f. -Moriche (Venezuela), Canangucha (Colombia) or Aguaje (Ecuador).
- *Oenocarpus* Mart.
- *Oenocarpus* bataua- Seje, Milpesos (Colombia), Milpé, Komaña (Uitoto), Ungurahua (Perú)
- *Orbignya* Mart. ex Endl.
- *Phytelephas* Cook
- *Phytelephas seemanii* - Tagua, Tagua palm. Ivory palm.
- *Phoenix*
- *Phoenix canariensis* Chabaud- Canary Island palm tree
- *Phoenix dactylifera* L. - date palm
- *Rhapis* L.f.°
- *Roystonea* O.F.Cook
- *Roystonea regia* H.B.K. - Royal palm, Cuban royal palm, Cuban royal palm
- *Roystoneaoleracea* O.F. Cook-Chaguaramo (Venezuela), royal palm (other countries)
- *Sabal* Adans.

- *Salacca* Reinw.
- *Syagrus*
- *Trachycarpus* L.
  - *Trachycarpus fortunei* (Hook.) H.Wendl.
- *Veitchia* H.Wendl. in Seem.
- *Wallichia* Roxb.
- *Washingtonia* H.Wendl.
  - *Washingtonia filifera* (Linden) H.Wendl.
  - *Washingtonia robusta*
- *Weittinia*

The following classification was proposed by N.W.Uhl and J.Dransfield in 1987 in *Genera palmarum: a classification based on the work of Harold E. Moore, Jr.* (but see new classifications such as Asmussen *et al.* 2006[20]):

The list of all genera of the botanical family Arecaceae in Appendix: Genera and tribes of Arecaceae.

Synonymy

Synonymy, according to APWeb[3] (visited January 2009):

- of Arecales: **Cocosales** Dumort. - **Arecanae** Takht. - **Arecidae** Takht. - **Phoenicopsida** Brongn.
- of Calamoideae: **Calamaceae** Perleb, **Lepidocaryaceae** O.F.Cook
- of Nypoideae: **Nypaceae** Le Maout & Decne.
- of Coryphoideae: **Borassaceae** Schultz Schultz Sch., **Coryphaceae** Schultz Sch., **Phoeniciaceae** Burnett, **Sabalaceae** Schultz Sch.
- of Ceroxyloideae: **Phytelephaceae** Perleb,
- of Arecoideae: **Acristaceae** O.F.Cook, **Ceroxylaceae** O.F.Cook, **Chamaedoraceae** O.F.Cook, **Cocosaceae** Schultz Sch., **Geonomataceae** O.F.Cook, **Iriarteaceae** O.F.Cook & Doyle, **Malortieaceae** O.F.Cook, **Manicariaceae** O.F.Cook, **Pseudophoeniciaceae** O.F.Cook, **Synechanthaceae** O.F.Cook

Economic importance

El Palmar National Park, Entre Ríos, Argentina, is an immense protected area of 8500 hectares with abundant specimens of yatay palm trees. The park attracts thousands of tourists annually.

In general, palms are one of the most important elements for Amazonian communities because of their economic, cultural and ecological value, since they provide food, housing and multiple items that satisfy their material needs. Many palm species have a great current and potential value as sources of food, oils, fibers, medicines and other products, including their value as ornamental plants; all of these potentialities (if used in a sustainable manner) can become a valuable source of resources for the economy.[27]

It is one of the most economically important botanical families.

Food plants come from *Areca*, *Attalea*, *Bactris*, *Cocos* (coconut palm, *Cocos nucifera*), *Elaeis* (oil-bearing, e.g. *Elaeis oleifera*), *Metroxylon* (providing starch), and *Phoenix* (date palm). Many genera have an edible apical bud. *Euterpe edulis* is the edible palmetto in the Southern Cone.

Other economically important palms are *Calamus* (and others called rattan), *Copernicia* (carnauba wax, *Copernicia cerifera*: wax palm), *Phytelephas* (tagua), *Raphia* (raffia), and many genera that provide thatch or fibers for tying.

Finally, the family includes a large number of ornamentals, *Caryota*, *Chamaerops*, *Livistona*, *Roystonea*, *Sabal*, *Syagrus*, *Washingtonia*, *Chamaedorea*, *Raphidophyllum*, *Thrinax*, *Coccothrinax*, *Licuala*, *Veitchia*, *Acoelorraphe*, *Butia*, *Copernicia*, *Dypsis*, and *Wodyetia*. Among the ornamentals are *Phoenix canariensis* (Canary Island palm) and *Roystonea regia* (Cuban royal palm) and many others.

Some species are cultivated over large areas, such as the coconut palm *Cocos nucifera*, the oil palm *Elaeis guineensis* and the date palm *Phoenix dactylifera.*

The sap of some species is concentrated or fermented to produce palm "honeys" and "wines".

The fruit of *Areca catechu* is chewed in Asia as a stimulant and is known as betel.

Disambiguation

Some species commonly called palms, or confused with them, are:

- *Cordyline australis*[28] (*Torbay palm, Ti palm,*[28] *Palm lily*[28]) (family Asparagaceae) and other representatives in the genus *Cordyline* and perhaps also in *Dracaena* with which *Cordyline* can be confused (and *Aloe* and *Agave*?).
- *Cycas revoluta* (*Sago palm*[28]) and the rest of the order Cycadales
- *Ravenala* (*Traveller's palm*[28]) (family Strelitziaceae)
- *Pandanus spiralis*, Screw palm,[28] and perhaps other *Pandanus* species.

spp.

- *Cyathea cunninghamii* (*Palm Fern*[28]) and other tree ferns (families Cyatheaceae and Dicksoniaceae) that can be confused with palm trees.
- *Setaria palmifolia* (*Palm grass*[28]), a Poaceae.
- *Carludovica palmata* (*Panama Hat Palm*[28]) and perhaps other members in the family Cyclanthaceae.

GENUS GAUSSIA

Gaussia

G. spirituana

Taxonomy	
Kingdom:	Plantae
Division:	Magnoliophyta
Class:	Liliopsida
Order:	Arecales
Family:	Arecaceae
Subfamily:	Arecoideae
Tribe:	Chamaedoreeae
Genus Gaussia: H.WENDL.	

Species

- *Gaussia attenuata*
- *Gaussia gomez-pompae*
- *Gaussia maya*
- *Gaussia princeps*
- *Gaussia spirituana*

Synonymy

- *Aeria* O.F.Cook (1901).
- *Opsiandra* O.F.Cook (1923).[1]

Gaussia is a genus of five species of flowering plants belonging to the palm family (Arecaceae). They are distributed in Mexico, Central America and the Antilles.

Description

They are trees with pinnate compound leaves. The species *G.attenuata* is found in Dominican Republic and Puerto Rico, *G.gomez-pompae* is found in Chiapas, Oaxaca and Veracruz in Mexico and *G.spirituana* is found in Cuba.

Taxonomy

The genus was described by Hermann Wendland and published in *Nachrichten von der Königlichen Gesellschaft der Wissenschaften und von der Georg-Augusts- Universität* 1865(14): 327-328. 1865.[2]

Etymology
Gaussia: generic name given in honor of the German mathematician Karl Friedrich Gauss (1777-1855).[3]

Species

- *Gaussia attenuata*
- *Gaussia gomez-pompae*
- *Gaussia maya*
- *Gaussia princeps*
- *Gaussia spirituana*

Gaussia spirituana

Gaussia spirituana

State of conservation

Endangered **(IUCN 2.3)**[1]

Taxonomy

Kingdom:	Plantae
(without rank):	Monocots
(without rank):	Commelinids
Order:	Arecales
Family:	Arecaceae
Genus:	*Gaussia*
Species:	***G. spirituana*** MOYA & LEIVA

[edit data in Wikidata] [edit data in Wikidata

Gaussia spirituana is a palm tree endemic to the Sierra de Jatibonico, in central-eastern Cuba.[2]

Description

Gaussia spirituana has stems are whitish in color, reaching a size of up to 7 meters in height and 30-35 centimeters in diameter, swollen at the base tapering upwards. The trees have up to ten pinnate compound leaves. The fruits are orange-red, 1 cm in diameter.[3]

The species is considered endangered based on the fact that only 150 individuals are known to be alive, fragmented into five subpopulations.[1] They are also threatened by habitat destruction and non-native pathogens.[1]

Taxonomy

Gaussia spirituana was described by Moya & Leiva and published in *Revista del Jardín Botánico Nacional* 12: 16. 1991 [1993].[4]

Etymology

Gaussia: generic name given in honor of the German mathematician Karl Friedrich Gauss (1777-1855).[5]

spirituana: Latin epithet

MATERIALS AND METHODS

Location of the study area

The study area is the Bamburanao massif, in north-central Cuba, which encompasses the landscape subdistrict of the plains, heights and small mountains of the mountain range (Acevedo, 1984).

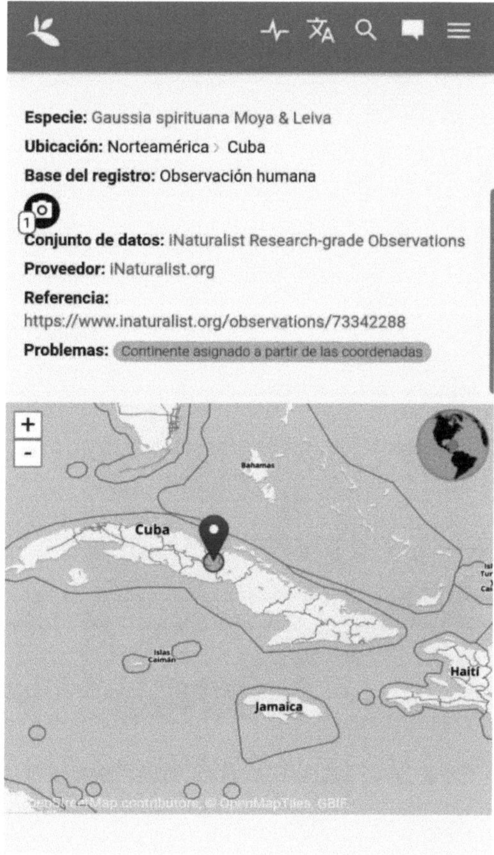

Gaussia spirituana Moya & Leiva

Publicado en: Revista Jard. Bot. Nac. Univ. Habana 12: 16 (1991 publ. 1993)
fuente: Catalogue of Life Checklist

9 REGISTROS

INFORMACIÓN GENERAL   ESTADÍSTICAS

1 REGISTRO CON IMÁGENES

5 REGISTROS GEOREFERENCIADOS

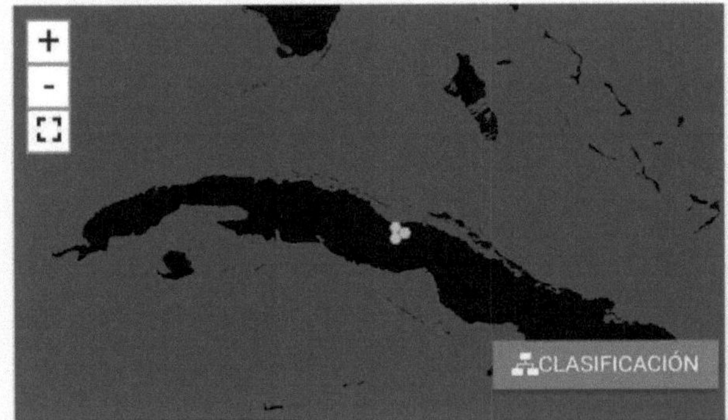

5 REGISTROS GEOREFERENCIADOS
500 km
Generado © OpenStreetMap contributors, © OpenMapTiles, GBIF.

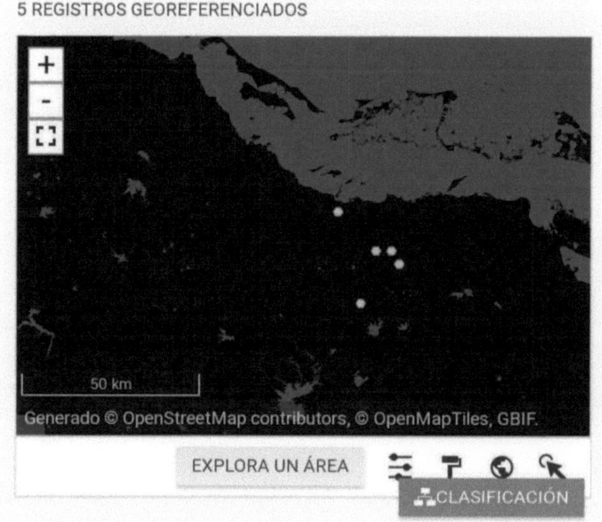
5 REGISTROS GEOREFERENCIADOS
50 km
Generado © OpenStreetMap contributors, © OpenMapTiles, GBIF.
EXPLORA UN ÁREA
CLASIFICACIÓN

Mountain massifs

Main Biological Diversity refuges in Cuba, considered Special Regions of Sustainable Development (REDS)

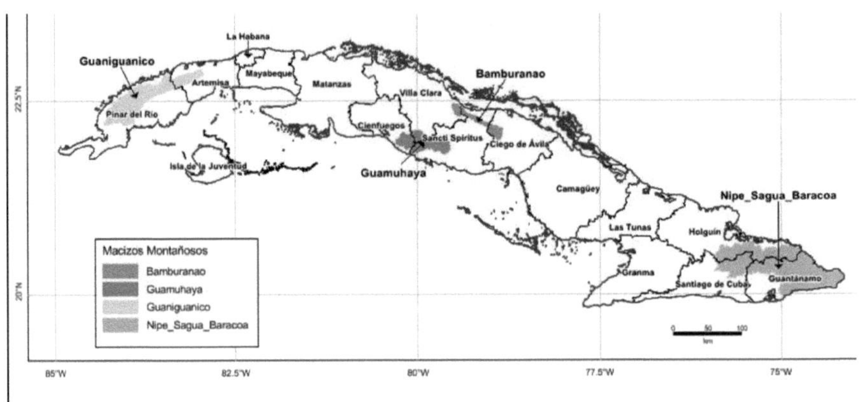

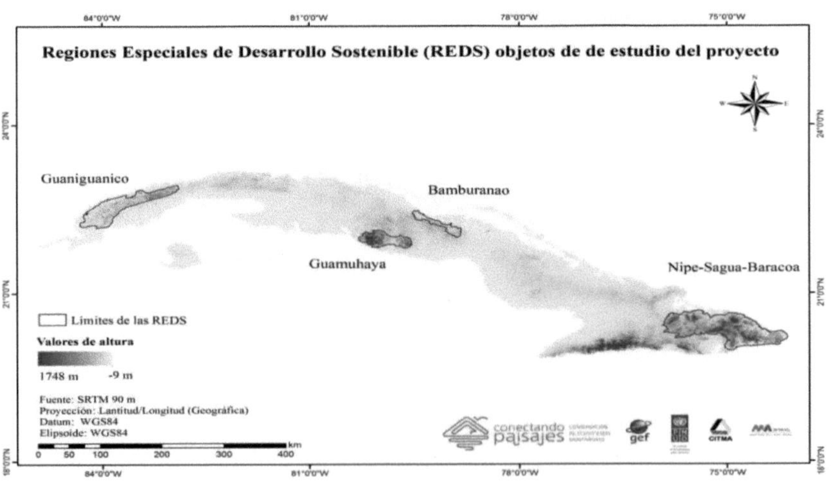

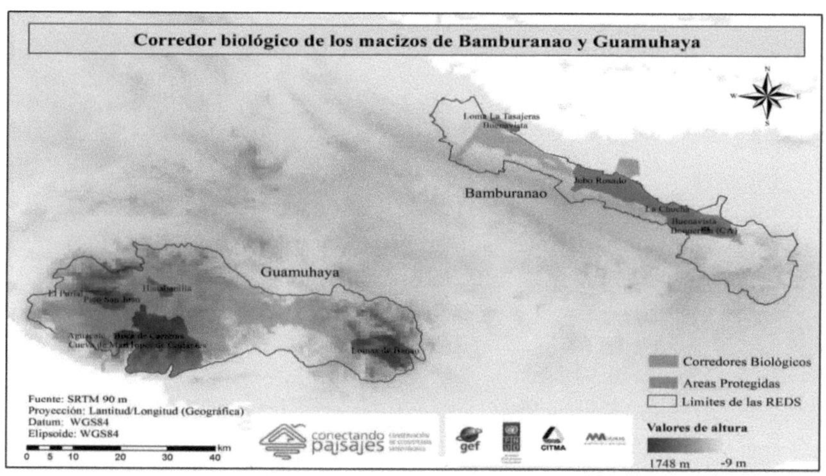

It occupies the eastern portion of the subdistrict of La Cordillera and extends from the Sagua la Chica River, which serves as a boundary with the central plains and heights of the north to the west, to the area of the plain from Júcaro to Morón or La Trocha, in the Camagüey-Maniabón district, along its eastern foothills, near the towns of Ciro " Redondo (Pina), Ceballos and Ciego de Ávila to the east. To the north it borders the northern central plain and the northwestern portion of the coastal plain from Júcaro to Morón and, to the south, along its southern foothills, with the plains, heights and small mountains of Cubanacán and with the southwestern portion of the plain from Júcaro to Morón. This group of landscapes is the most extensive of the subdistrict with about 2,040 *km2 (*Figure 1).

Due to its structure, it also forms part of the articulation zone between the myogeosyncline and the eugeosyncline (Camajuaní and Placetas areas) and also extends to part of the former Santa Clara anticlinorium of the eugeosyncline (Santa Clara area). It is an area of overthrusts and numerous faults, such as the *Las Villas* fault *(*fig. 6.4).

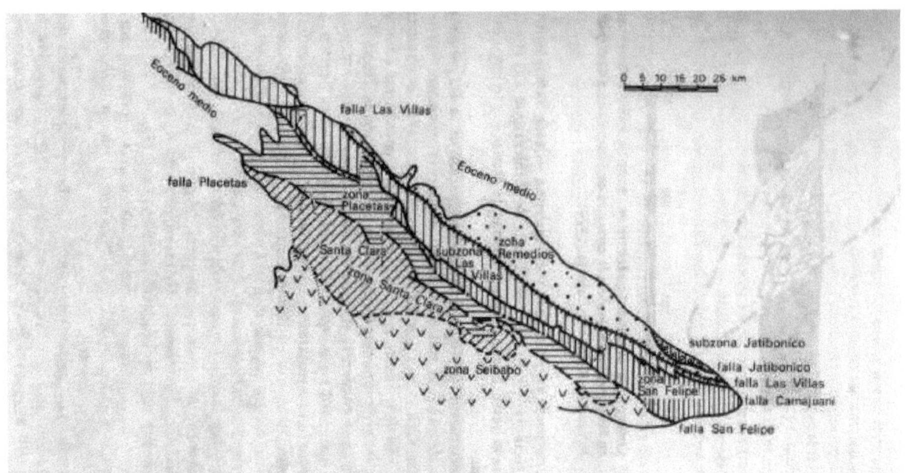

Fig. 6.6 Zonas estructuro-faciales de la región de Santa Clara delimitadas por fallas alineadas de noroeste a sureste (según Kniper y Cabrera, 1974).

The rocks that constitute it are similar to those of the rest of the subdistrict, in essence, limestone and effusive-sedimentary of the Cretaceous and ultrabasic and basic. The relief, in general, is of small mountains whose heights do not exceed 500 *m* and denudational and karstic plains: to the north it is of small petrogenic heights with fragments of abrasive terraces and monadnocks; Towards the center of the landscape there are chains of residual heights, among which the Bamburanao mountain range (347 *m)* and a denudation plains are the most outstanding. Towards the west, where the relief becomes more complex, there are the Meneses (305 m) and Cueto (240 m) mountain ranges, the hills of La Canoa (250 m), of strongly karstic relief and the Jatibonico mountain range (443 m), which constitutes the culminating point of the subdistrict.

The Jatibonico mountain range is formed by small mountains of karst erosivotectonic blocks, crossed by the Jatibonico river from the north with a portion of its subterranean course, which present several flattening surfaces, one especially notable, located at an altitude of about 200 *m*, which is very well expressed in Las Llanadas, the Alunado valley and in a plain located at the foot of the Mabuya hills, near Boquerón.

The subway riverbed of the northern Jatibonico or Boquerón river was crossed for the first time in 1961 by a Polish-Cuban speleological expedition.

This riverbed was excavated along a fault and, inside it, there is a deep lake about 350 *m* long; at a higher level is the Bonita de Boquerones cave, a cavity that is currently hydrologically inactive (Acevedo, 1965).

The soils are distributed from northwest to southeast in the following order, humified limestone, latosolic (red ferrallitic), red limestone (rendzinas) and tropical brown soils.

The territory is, in general, more humid than the previous one and receives an average annual rainfall of about 1,700 *mm,* although it decreases towards the bordering areas to about 1,400 *mm,* therefore most of the area corresponds to the landscapes of the climatic zone of temporarily humid tropical forests, with 70 to 80 % of annual rainfall in the rainy period.

The vegetation, in almost all the group of landscapes, is of sandy and clayey savannahs, with planifolia trees and, in the low mountains and stony mounds, it is of forests and complex vegetation of mogotes: arboreal and shrubby, The main rivers of the zone are: Camajuaní, a tributary of the Sagua la Chica; Manacas; Caonao, a tributary of the Zaza; Jatibonico del Norte and Jatibonico del Sur, which are born in this group of landscapes and flow into the north and south coast respectively and, finally, the Chambas.

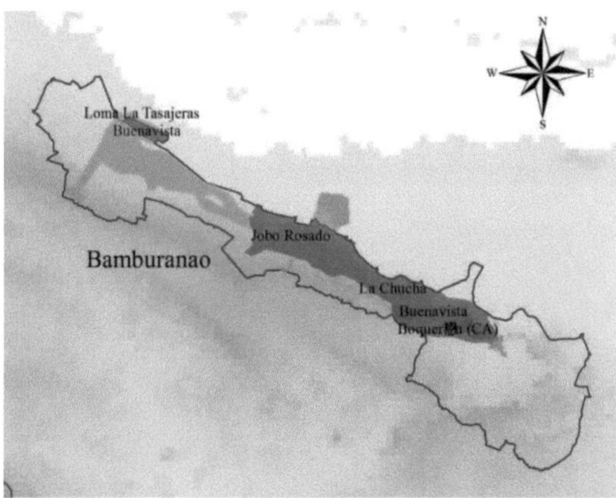

Figure 1. Bamburanao Mountain Massif, areal distribution of relict populations of *G. spirituana*.

In this landscape context are the five populations of *Gaussia espirituana*, a species distributed in the hills of Canoa and the Jatibonico mountain range, which from a political-administrative point of view cover portions of the municipalities of Yaguajay and Florencia, belonging to the provinces of Sancti Spíritus and Ciego de Ávila (Figure 2).

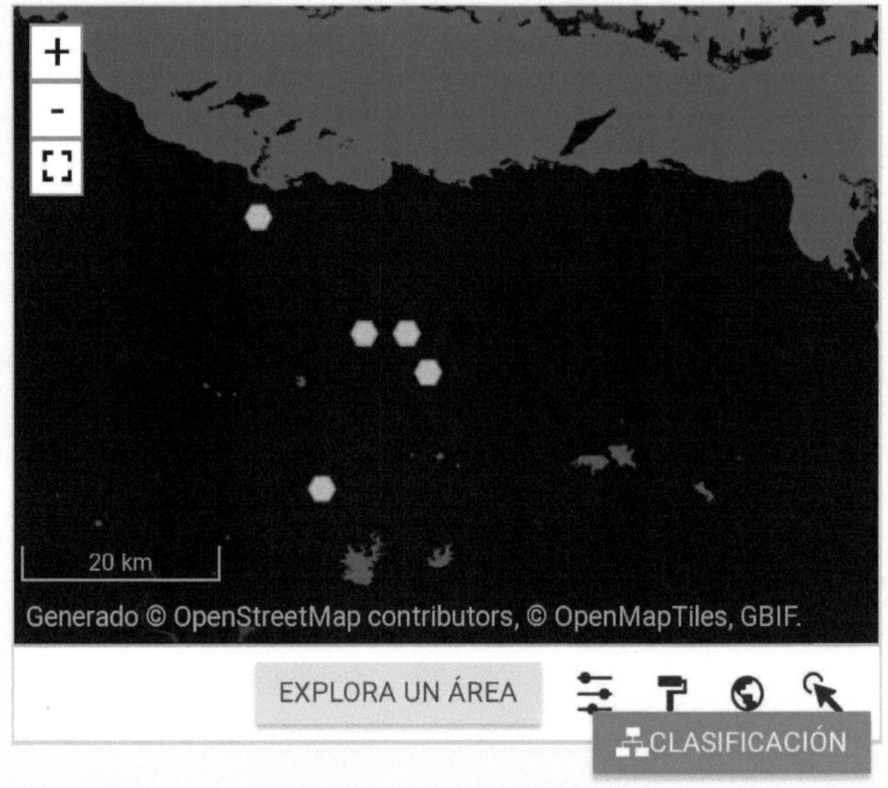

Figure 2. Locations of the five populations of *G. spirituana*.

Individual of G. spirituana spotted towards the ornamental area of the Botanical Garden of Sancti Spíritus, Cuba.

Specific methodology used in the present study

In many cases, plant populations are important sources of resources for agriculture, forestry, recreation and wildlife refuge, so their study is a necessary aspect to address.

Within the characteristics of plant populations in general, the density and distribution in those plant populations are important because in forests they are the most necessary indicators to evaluate the possibility of biomass production and its extraction, thus, the density informs us of the number of individuals present in a given area at a given time and the distribution in the occupied space.

Twenty-five plots were chosen at random, five for each population of the Sierra Espirituana palm, and we proceeded as follows:

a) Density calculation

To determine the relationship between the number of individuals and the space they occupy, a specific area or sample size of the *G. spirituana* population of 100 m^2 (one hectare) was measured.

Density = No. of individuals
Area

Five samples were taken for each population to obtain the average.

b) Determination of the distribution of individuals in the population Under natural conditions, plants are not distributed in space in the same way and there are three broad patterns: random, uniform and grouped.

- Random: individuals are randomly distributed, this distribution is very rare in nature, in conditions of a very uniform environment.
- Uniform: individuals are ordered.
- In groups: individuals are grouped in variable densities from one area to another.

To determine the distribution in the plant population, the mean variance ratio (c) was obtained using the formula:

C =S^2 , where: S^2 - variance and X- mean

X

The number of individuals in different zones in the indicated area was taken from the population studied: the S^2 was found for these data and C was obtained if:

C = 1, the distribution is random,

C > 1, the distribution is clustered, C < 1, the distribution is uniform.

c) Vegetation profile

Many plant studies are accompanied by vegetation profiles, the most accurate is to have a representative length of the vegetation under study. In Cuba it is frequently used between 25 - 50 m linear (*e.g.* Capote and Berazaín, 1984; González-Torres and Berazaín, 2004). As for the width or depth of this linear area, it is frequently 1 m (*e. g.* Matos and Torres, 2000; González-Torres and Berazaín, 2004). In general, a vegetation profile should represent, to scale, the height and distance between plants, so the values of the horizontal and vertical axes should not be missing. Each species in the profile should be identified with some abbreviation, for which the initial of the genus and the species that make up its scientific name are frequently used, which are referred to in a legend.

Figure 4. Several specimens of *G. spirituana* in their natural environment.

RESULTS AND DISCUSSION

The first results obtained from the demographic study of the *G. spirituana* populations refer to the location and georeferencing of the five demos (Table 1).

Table 1. Location and georeferencing of the populations of *G. spirituana* endemic to Cuba. Map scale 1:250 000, Lambert coordinates.

PROVINCE	MUNICIPALITY	TERRITORY	LOCATION	COORDE-NADAS-x	COORDE-NADAS-and
Ciego de Avila	Florence	Jatibonico Mountains	Anchovy	17-8	26-76
Ciego de Avila	Florence	Jatibonico Mountains	Jagueycito	17-8	25-71
Sancti Spíritus	Yaguajay	Lomas de la Canoa	La Chucha	17-8	27-68
Sancti Spíritus	Yaguajay	Sierra de Meneses	Chinese Stone	17-8	27-68
Sancti Spíritus	Yaguajay	Lomas de la Canoa	South of Mayajigua	17-12	23-68

Of the five towns, three are located in the province of Sancti Spiritus and two in the province of Ciego de Avila. The three demos are located in the municipality of Yaguajay, while the other two are in the municipality of Ciego de Avila. Two are also located in the Jatibonico mountain range, another two in Lomas de la Canoa and the remaining one in the Meneses mountain range.

The most numerous populations in terms of individuals are those belonging to the territory of Avila, but the species was discovered, studied and made known to the scientific community by the discovery made in Piedra China in 1989 and published in the journal of the National Botanical Garden in 1991.

DEMOGRAPHY OF THE SPECIES

The most numerous populations are those of Jagueycito (53) and Boquerón (48), both in the Jatibonico mountain range, for a total of 101 individuals on Avileño properties, while the Espiritu demos total 106 specimens, especially in the hills surrounding the town of Mayajigua (Table 2).

Table 2. Densities by localities and total of endemic *G. spirituana* endemic to Cuba.

LOCATION	DENSITY
Anchovy	48
Jagueycito	53
La Chucha	36
Chinese Stone	31
South of Mayajigua	39
T O T A L E S	207

In relation to the age structure of the populations of the species, it can be said that adult individuals predominate numerically (147), while juveniles (46) and seedlings (14) are present in smaller numbers, for a grand total of 207 specimens of the Sierra Espirituana palm (Table 3).

Table 3. Age composition of *G. spirituana* populations.

LOCATION	ADULTS	YOUTH	PLANTULAS	TOTAL
Anchovy	33	12	3	48
Jagueycito	37	13	3	53
La Chucha	24	9	3	36
Chinese Stone	21	6	4	31
South of Mayajigua	32	6	1	39
TOTALES	147	46	14	207

Regarding the measures of central tendency obtained by means of the more general statistics shown in Table 4, it can be said that these provided the values of the mean and variance necessary to calculate C, thus determining the values necessary to define the type of distribution of the species in its distribution areal.

Values of the most general statistics for the distribution of the population of *G. spirituana*, endemic and threatened palm of central Cuba.

Sample	Mean Individuals	Variance	Standard Deviation	Standard Error	Total Individuals
Sample 1	19,55	756,605	27,506	19,45	39,1
Sample 2	20,55	836,405	28,921	20,45	41,1
Sample 3	20,06	795,207	28,199	19,94	40,12
Sample 4	21,05	877,805	29,628	20,95	42,1
Sample 5	24,055	1146,726	33,863	23,945	48,11
Sample 6	26,55	1399,205	37,406	26,45	53,1
Sample 7	25,05	1245,005	35,285	24,95	50,1
Sample 8	25,55	1295,405	35,992	25,45	51,1
Sample 9	25,05	1245,005	35,285	24,95	50,1
Sample 10	26,055	1346,286	36,692	25,945	52,11
Sample 11	17,05	574,605	23,971	16,95	34,1
Sample 12	17,55	609,005	24,678	17,45	35,1
Sample 13	18,05	644,405	25,385	17,95	36,1
Sample 14	15,56	476,787	21,835	15,44	31,12
Sample 15	16,055	508,486	22,55	15,945	32,11
Sample 16	14,555	417,316	20,428	14,445	29,11
Sample 17	15,55	477,405	21,85	15,45	31,1
Sample 18	15,05	447,005	21,142	14,95	30,1
Sample 19	14,055	388,926	19,721	13,945	28,11
Sample 20	15,56	476,787	21,835	15,44	31,12
Sample 21	18,55	680,805	26,092	18,45	37,1
Sample 22	18,05	644,405	25,385	17,95	36,1
Sample 23	19,55	756,605	27,506	19,45	39,1
Sample 24	17,565	607,958	24,657	17,435	35,13
Sample 25	19,05	718,205	26,799	18,95	38,1

The calculation of C, allowed us to know the type of distribution of the five populations of *G. spirituana*, observing that in the 25 sampling plots C data were obtained well above 1, therefore the distribution of the species in all cases is grouped (Table 5).

Table 5. C values calculated to determine the type of distribution of the species.

PLOTS	C
Plot No.1	38,70
Plot No.2	40,70
Plot No.3	39,64
Plot No.4	41,70
Plot No.5	47,67
Plot No.6	52,70
Plot No.7	49,70
Plot No.8	30,70
Plot No.9	49,70
Plot No.10	51,67
Plot No.11	33,70
Plot No.12	34,70
Plot No.13	35,70
Plot No.14	36,64
Plot No.15	31,67
Plot No.16	28,67
Plot No.17	36,70
Plot No.18	29,70
Plot No.19	27,67
Plot No.20	30,64
Plot No.21	36,70
Plot No.22	35,70
Plot No.23	38,70
Plot No.24	24,66
Plot No.25	37,70

This result is crucial because the grouped distribution classifies as a very successful survival strategy, since the grouping of individuals or their organization in groups provides great advantages for the resilience of the species in question, since it facilitates reproduction due to the fact that the proximity between specimens favors the exchange of gametes and cross-fertilization, as well as protection against meteorological events by acting as a windbreak. The results of this grouping can be seen in the number of seedlings and juveniles observed in the sampling units. This fact allows us to predict the sustainability of the population size of this species, which has been declared by the IUCN as endangered.

The vegetation profile is shown in Figure 5.

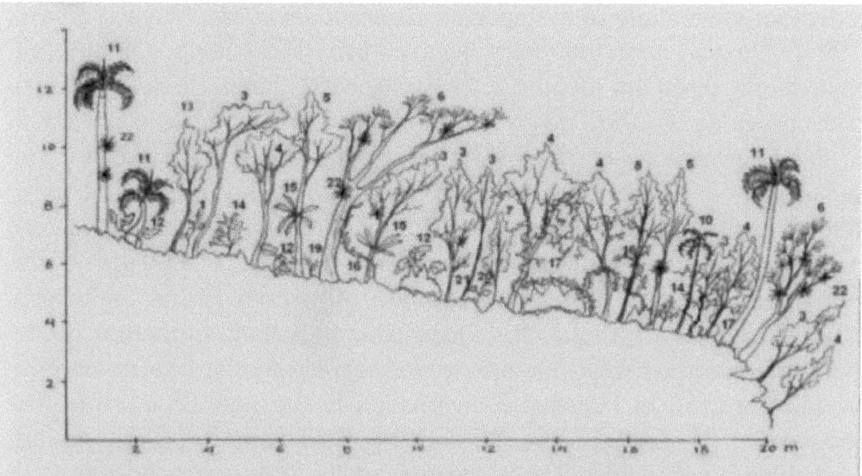

Fig. 5. Complejo de vegetación de mogotes, Cima del Farallón de Agua Santa: 1- *Chascotheca neopeltandra (Griseb.) Urb., 2- Smilax domingensis Willd., 3- Savia sessiliflora (Sw.) Willd., 4- Gymnanthes lucida Sw., 5- Cordia gerascanthus L., 6- Plumeria obtusa L., 7- Picramnia pentandra Sw., 8- Citharexylum spinosum L., 9- Erythroxylum havanense Jacq. var. havanense, 10- Comocladia platyphylla A. Rich. ex Griseb., 11- Gaussia spirituana Moya & Leiva, 12- Philodendron lacerum (Jacq.) Schott, 13- Jacaranda caerulea (L.) Juss., 14- Erythroxylum havanense Jacq. var.*

This figure shows how *G. spirituana* lives in association with the phytocenosis called the mogotes vegetation complex, a plant formation present in the Saguense phytogeographic district and throughout much of the Bamburanao Mountain Massif.

In this type of scrub, the common ancestor of *G. princeps* and *G. spirituana* evolved independently through the evolutionary force of speciation until the emergence of these two endemic and endangered Cuban species.

In the mogotiform karst domes of Bamburanao, this species survives in relict populations, forming part of the plant biocenosis known as mogotes vegetation.

CONCLUSIONS

A demographic study of the species *Gaussia spirituana* Moya *et* Leiva, (1991), showed that the palm lives in five populations. Of the five populations, three are located in the province of Sancti Spiritus and two in the province of Ciego de Avila. The three Espiritu demos are located in the municipality of Yaguajay, while the other two are in the municipality of Ciego de Avila. Two are also located in the Jatibonico mountain range, another two in the Canoa hills and the remaining one in the Meneses mountain range. The most numerous populations are those of Jagueycito (53) and Boquerón (48), both in the Jatibonico mountain range, for a total of 101 individuals on Avileño properties, while the Espiritu demos total 106 specimens, especially the one in the hills around the town of Mayajigua. In relation to the age structure of the populations of the species, it can be said that adult individuals predominate numerically (147), while juveniles (46) and seedlings (14) are present in smaller numbers, for a grand total of 207 specimens of the Sierra Espirituana palm.
The calculation of C, allowed us to know the type of distribution of the five populations of *G. spirituana*, observing that in the 25 sampling plots C data was obtained well above 1, therefore the distribution of the species in all cases is grouped. This result is crucial because the grouped distribution is classified as a very successful survival strategy, since the grouping of individuals or their organization in groups provides great advantages for the resilience of the species in question, since it facilitates reproduction due to the fact that the proximity between specimens favors the exchange of gametes and cross-fertilization, as well as protection against meteorological events by acting as a windbreak.
The results of this grouping are observed in the number of seedlings and juveniles observed in the sampling units, which allows predicting the sustainability of the population size of this species that has been declared by the IUCN as endangered. The vegetation profile allows inferring that *G. spirituana* lives in association with the phytocenosis known as the mogotes vegetation complex, a plant formation present in the Saguense phytogeographic district and throughout a large part of the Bamburanao Mountain Massif.

RECOMMENDATIONS

To submit this research to the authorities of the National Center of Protected Areas of Cuba, so that the creation of a protected area that includes the entire distribution areal of this endangered endemic species of the flora of the island of Ciuba is prioritized.

REFERENCES

1. 1 2 Elspeth Haston, James E. Richardson, Peter F. Stevens, Mark W. Chase, David J. Harris. The Linear Angiosperm Phylogeny Group (LAPG) III: a linear sequence of the families in APG III Botanical Journal of the Linnean Society, Vol. 161, No. 2. (2009), pp. 128-131. doi:10.1111/j.1095-8339.2009.01000.x Key: citeulike:6006207 pdf: http://onlinelibrary.wiley.com/doi/10.1111/j.1095-8339.2009.01000.x/pdf
2. 1 2 The Angiosperm Phylogeny Group III ("APG III", in alphabetical order: Brigitta Bremer, Kåre Bremer, Mark W. Chase, Michael F. Fay, James L. Reveal, Douglas E. Soltis, Pamela S. Soltis, and Peter F. Stevens, also contributed by Arne A. Anderberg, Michael J. Moore, Richard G. Olmstead, Paula J. Rudall, Kenneth J. Sytsma, David C. Tank, Kenneth Wurdack, Jenny Q.-Y. Xiang, and Sue Zmarzty) (2009). "An update of the Angiosperm Phylogeny Group classification for the orders and families of flowering plants: APG III." (pdf). *Botanical Journal of the Linnean Society* (161): 105-121.
3. 1 2 Stevens, P. F. (2001 onwards). "Angiosperm Phylogeny Website (Version 9, June 2008, and updated since)". Accessed 12 January 2009.
4. ↑ {video{quote web|url=https://www.youtube.com/watch?v=YSP40xXRNzQ&feature=youtu.be}}
5. 1 2 José Antonio del Cañizo (2002). *Palmeras*. Ediciones Mundi-Prensa. ISBN 84-7114-989-3.
6. ↑ Izco, J., Barreno, E. (1997). *botany*. McGraw_Hill. ISBN 84-486-0182-3.
7. ↑ Galeano, G.2000. Conservation status of palms in Colombia. In: Revista Pérez - Arbelaezia, Volume 5, Nº 11 (April); Pp. 68 - 70.
8. ↑ Henderson, A. (1986). "A review of pollination studies in the palms.".
Bot. Rev. (52): 221-259.
9. ↑ Zona, S.; Henderson, A. (1989). "A review of animal-mediated seed dispersal in palms.". *Selbyana* (11): 6-21.
10. ↑ Dransfield, J. (1986). "A guide to collecting palms". *Ann. Missouri Bot. Gard.* (73): 166-176.
11. ↑ Dransfield, J.; Uhl, N. W. (1998). "Palmae.". In K. Kubitzki. *The families and genera of vascular plants. vol. 4, Monocotyledons:*

Alismatanae and Commelinanae (except Gramineae). Berlin: Springer-Verlag. pp. 306-389.
12. ↑ Henderson, A. (1995). *The palms of the Amazon*. New York: Oxford University Press.
13. ↑ Henderson, A.; Galeano, G., and Bernal, R. (1995). *Field guide to the palms of the Americas*. Princeton, NJ: Princeton University Press.
14. ↑ Moore, H. E. (1973). "The major groups of palms and their distribution.". *Gentes Herb.* (11): 27-141.
15. ↑ Moore, H. E.; Uhl, N. W. (1982). "The major trends of evolution in palms.". *Bot. Rev.* (48): 1-69.
16. ↑ Tomlinson, P. B. (1990). *The structural biology of palms*. Oxford: Clarendon Press.
17. ↑ Uhl, N. W.; Dransfield, J. (1987). *Genera palmarum*. Ithaca, NY: L. H. Bailey Hortorium and International Palm Society.
18. ↑ Zona, S. (1997). "The genera of Palmae (Arecaceae) in the Southeastern United States.". *Harvard Pap. Bot.* (11): 71-107.
19. ↑ Dransfield, J.; Uhl, N., Asmussen, C. B., Baker, W. J., Harley, M. M., and Lewis, C. E. (2005). "A new phylogenetic classification of the palm family, Arecaceae.". *Kew Bull.* (60): 559-569.
20. 1 2 Asmussen, C. B.; Dransfield, J., Deichmann, V., Barfod, A. S., Pintaud, J. -C., and Baker, W. J. (2006). "A new subfamily classification of the palm family (Arecaceae): evidence from plastid DNA phylogeny.". *Bot. j. Linnean Soc.* (151): 15-38. Accessed February 25, 2008.
21. 1 2 Uhl, N. W.; Dransfield, J., Davis, J. I., Luckow, M. A., Hansen, K. H., and Doyle, J. J. (1995). "Phylogenetic relationships among palms: cladistic analyses of morphological and chloroplast DNA restriction site variation.". In Rudall, P. J., Cribb, P. J., Cutler, D. F., and Humphries, C.
J. *Monocotyledons: Systematics and evolution* (Royal Botanic Gardens ed.). Kew. pp. 623-661.
22. 1 2 Hahn, W. J. (2002). "A molecular phylogenetic study of the Palmae (Arecaceae) based on *atpB, rucL,* and 18S nrDNA sequences.". *Syst. Biol.* (51): 92-112.
23. ↑ Baker, W. J.; Zona, S., Heatubun, C. D., Lewis, C. E., Maturbongs, R. A., and Norup, M. V. (2006). "*Dransfieldia* (Arecaceae): a new palm genus from western New Guinea.". *Syst. Bot.* (31): 61-69.
24. ↑ Asmussen, C. B.; Baker, W. J., and Dransfield, J. (2000).

"Phylogeny of the palm family (Arecaceae) based on *rps16* intron and *trnL-trnF* plastid DNA sequences.". In Wilson, K. L. and Morrison, D. A. *Monocots: Systematics and evolution.* (CSIRO Publ. ed.). Collingwood, Australia. pp. 525-535.

25. ↑ Asmussen, C. B.; Chase, M. W. (2001). "Encoding and noncoding plastid DNA in palm systematics.". *Amer. J. Bot.* (88): 1103-1117. Accessed February 25, 2008.

26. ↑ APG II (2003). "An Update of the Angiosperm Phylogeny Group Classification for the orders and families of flowering plants: APG II." (pdf). *Botanical Journal of the Linnean Society (*141): 399-436. Accessed January 12, 2009.

27. ↑ Galeano, G. 1992. The palms of the Araracuara region. Studies in the Colombian Amazon. Facultad de Ciencias Naturales, Universidad Nacional De Colombia. Second Edition. Bogotá, Colombia. 179 Pág.

28. 1 2 3 4 5 6 7 8 9 (c) FAO 1995. *Tropical Palms*. Introduction. http://www.fao.org/docrep/X0451E/X0451e03.htm *NON-WOOD FOREST PRODUCTS* 10. FAO - Food and Agriculture Organization of the United Nations. ISBN 92-5-104213-6

REFERENCES ON THE SPECIES

1. 1 2 3 Moya, C. (1998). **'Gaussia spirituana'**. *IUCN Red List* 2007. IUCN 2007. Accessed 16 November 2007.

2. ↑ "*Gaussia spirituana*". *Royal Botanic Gardens, Kew: World Checklist of Selected Plant Families*. Accessed December 8, 2006. (broken link available from Internet Archive; see history and latest version).

3. ↑ Henderson, Andrew; Gloria Galeano; Rodrigo Bernal (1995). *Field Guide to the Palms of the Americas*. Princeton, New Jersey: Princeton University Press. ISBN 0-691-08537-4.

4. ↑ "*Gaussia spirituana*". *Tropicos.org. Missouri Botanical Garden*. Accessed March 1, 2015.

5. ↑ (J. Dransfield, N. Uhl, C. Asmussen, W.J. Baker, M. Harley and C. Lewis. 2008)

OTHER SPECIES REFERENCES

References

1. ↑ "*Gaussia*". *Royal Botanic Gardens, Kew: World Checklist of Selected Plant Families*. Accessed August 6, 2009.
2. ↑ "*Gaussia*". *Tropicos.org. Missouri Botanical Garden*. Accessed February 5, 2013.
3. ↑ (J. Dransfield, N. Uhl, C. Asmussen, W.J. Baker, M. Harley and C. Lewis. 2008)

- "*Gaussia* H.Wendl., Nachr. Königl. Ges. Wiss. Georg-Augusts-Univ. 1865: 327 (1865)." *Royal Botanic Gardens, Kew: World Checklist of Selected Plant Families*. Accessed 12 December 2006 (broken link available from Internet Archive; see history and latest version).
- George Proctor. 2005. Arecaceae (Palmae). Pp. 135-153 *in* Pedro Acevedo-Rodríguez and Mark T. Strong. Monocots and Gymnosperms of Puerto Rico and the Virgin Islands. *Contributions from the United States National Herbarium* Volume 52.

LITERATURE CITED

Acevedo-Rodríguez, P. and M. T. Strong. 2010. Catalogue of Seed Plants of the West Indies. *Smithsonian Contributions to Botany* 98: 1-1192.

Agrawal, S., P. Joshi, Y. Shukla and Roy P. 2003. Spot-Vegetation multi temporal data for classifying vegetation in South Central Asia. *Curr. Sci. India* 84 (11): 1440-1448.

Alain, Br. 1953. Flora of Cuba 3. Dicotyledons: Malpighiaceae to Myrtaceae. *Occasional* Contributions *of the Museum of Natural History Colegio "De La Salle"* 13: 1-472.

Alain, Br. 1957. Flora of Cuba 4. Dicotyledons: Melastomataceae to Plantaginaceae. *Occasional Contributions of the Museo de Historia Natural Colegio "De La Salle"* 16: 1-556.

Alain, Br. 1964. *Flora of Cuba 5. Dicotyledons: Rubiaceae to Compositae.*

Association of Students of Biological Sciences, Havana, 320 pp.

Alain, Br. 1974. *Flora of Cuba. Supplement.* Cuban Book Institute. Havana, 150 pp.

Alanís, E. 2010. Natural regeneration and post-fire ecological restoration of a mixed forest in Chipinque Ecological Park, Mexico. [Unpublished] PhD Thesis. Faculty of Forestry Sciences, Universidad Autónoma de Nuevo León, 119 pp.

Aymard, G., J. F. Quin, M. Rugiero, and G.S. Waggoner. 1995. The 0.1 Hectare methodology: a method for rapid assessment of woody plant diversity. *Handout* 7(1): 1-16.

Báez, S. A., L. Pintado and F. Hernández. 2016. Relationship between birds and dendrometric variables in plantations of *Pinus caribaea* Morelet var. *caribaea.*

W. H. Barret et Golfari in Viñales, Cuba *Revista Mexicana de Ciencias Forestales* 7(33): 8-19.

Barbour, M. G., J. H. Burk, and W. D. Pitts. 1987. *Terrestrial Plant Ecology.* Secound Edition. The- Benjamin/Cummings Publishing Company, California, 634 pp.

Barrios, D. 2008. Pollination biology of *Leptocereus scopulophilus* (Cactaceae) in Pan de Matanzas, Cuba. Bachelor's thesis. Faculty of Biology, University of Havana, 44 pp.

Barrios, D. 2012. Population structure, dispersal and recruitment microsites of *Leptocereus scopulophilus* (Cactaceae), Cuba. Master's thesis. National Botanical Garden of Cuba, University of Havana, 51 pp.

Berazaín, R. 2008. Update of the list of endemic genera of spermatophytes. *Journal of the National Botanical Garden* 29 (3):3-10.

Berazaín, R., F. Areces, J. Lazcano and L. R. González- Torres. 2005. *Red list of Cuban vascular flora*. Documents Atlantic Botanical Garden, Gijón, 86 pp.

Betancourt, J., A. Paz, E. Diaz and M. Faife. 2015. *Melocactus guitartii* León: the conservation of an endemic of the Spíritus flora. *Flora y fauna* 19 (1): 17 20.

Borhidi, A. 1991. *Phytogeography and vegetation ecology of Cuba*. Akadémiai Kiadó, Budapest. 858 pp.

Braun-Blanquet, J. 1964. *Pflazensociologie*. SpringerVerlag, Vienne-New York, 885 pp.

Bullock, J. M. 2006. Plants. Pp. 186-213. In: *Ecological Census Techniques: a handbook* (W. J. Sutherland, Ed.) Cambridge University Press. New York, 432 pp.

Capote, R. P. and R. Berazaín. 1984. Classification of the plant formations of Cuba. *Revista Jardín Botánico Nacional* 5 (2): 27-75.

Capote, R. P., E. E. García, and C. Sánchez. Sanchez. 1983. The vegetation of the Sierra del Rosario Ecological Station. *Journal of the National Botanical Garden* 4 (2): 97-143.

Duivenvoorden, J. F. 1994. Vascular plant species counts in the rainforest of middle Caquetá area, Colombian Amazonian. *Biodiversity and Conservation* 3: 685-715.

Elzinga, C. L., D. W. Salzer, and J. W. Willoughby. 1998. *Measuring and monitoring plant populations*. Bureau of Land Management, California, 477 pp.

Esparza-Olguín, L.; Valverde, T. and E. Valchis-Anaya. 2002. Demographic analysis of a rare columnar cactus (*Neobuxbaumia macrocephala*) in the Tehuacan Valley, Mexico. *Biological Conservation* 103: 349-359.

Fairbanks, D. and K. Mcwire. 2004. Patterns of floristic richness in vegetation communities of California: Regional scale analysis with multi-temporal NDVI. *Global Ecology and Biogeography* 13 (3): 221-235.

Falcón, A., J. P García-Lahera, and N. V. Hernández. 2016. New localities for *Maxonia apiifolia* (*Dryopteridaceae*) in Sancti Spíritus, Cuba. *Cuban Journal of Biological Sciences* 5 (1): 23-31.

Feeley, K. J., T. Gillespie, and J. Terborgh. 2005. The utility of spectral indices from lysat ETM+ for measuring the structure and composition of tropical dry forests. *Biotropica* 37 (4): 508-519.

Feinsinger, P. 2004. *The design of field studies for biodiversity conservation*. Editorial FAN, Santa Cruz de la Sierra, 242 pp.

Finol, H. 1971. New parameters to be considered in the structural analysis of tropical virgin forests. *Revista Forestal Venezolana* 14 (21): 29- 42.

Garcıa-Beltrá n, J. A., J. L. Fiallo, N. Esquivel, K. Meirama, I. Rodrıguez, B. Falcó n, V. Pé rez, and L. R. Gonzá lez-Torres. 2016. Effect of fire on the population structure of *Hypericum styphelioides* subsp. *styphelioides* (Hypericaceae) in the Reserva Ecoló gica "Los Pretiles", Cuba. *Journal of the National Botanical Garden of the University of Havana* 37: 19-27.

Gatsuk, L. E., O. V. Smirnova, L. I. Vorontzova, L. B. Zauglnova, and L. A. Zhukova. 1980. Age stages of plants of various growth forms: a review. *Journal of Ecology* 68: 675- 696.

Gentry, A. H. 1992. Tropical forest biodiversity: distributional patterns and their conservational significance. *Oikos* 63:19-28.

Godínez-Álvarez, H., J. E. Herrick, M. Mattocks, D. Toledo, and J. Van Zee. 2009. Comparison on three vegetation monitoring methods: their relative utility for ecological assessment and monitoring. *Ecological Indicators* 9: 1001- 1008.

Gómez-Hechevarría, J. L. 2016. *Spirotecoma holguinensis* a species to consider in ecological restoration. *Bissea* 10 (special issue 1):76.

Gómez-Hechavarría, J. L. and N. Cuellar. 2011-2012. Flora of the serpentinites of San Andres, Holguin, Cuba *Journal of the National Botanical Garden* 32-.
33: 111-124.

González-Gutiérrez, P. A., J. L. Gómez-Hechavarría, O. Leyva y Y. Hernández. 2015. Flora of the Reserva Florística Manejada cabo Lucrecia- punta de Mulas, Banes, Holguín. *Journal of the National Botanical Garden* 36: 65-77.

González-Gutierrez, P. A., S. I. Suárez, S. Sigarreta, A. Fernández and

O. Laffita. 2004-2005 Flora and vegetation of Caletica, Rafael Freyre, Holguín. *Journal of the National Botanical Garden* 25/26: 131-140.
González-Oliva, L., M. A. Gutiérrez, A. Urquiola and A. Urquiola. 2004. Spatial pattern of *Buxus wrightii* Muell. Arg., an endemic species of the ultramafic region of Cajálbana. Pp. 55-56. In *Ultramafic Rocks: Their soils, Vegetation and Fauna* (Boyd R., Baker A. and J. Proctor, Eds.)Science Reviews, St Albans. UK.

González-Oliva, L. 2010. *Population ecology and life history traits of the endemic species Amaranthus minimus* (Amaranthaceae)*:* implications for its conservation. PhD Thesis. Faculty of Biology, University of Havana, 121 pp.
González-Oliva, L., L. R. González-Torres, A. Palmarola and D. Barrios (Eds.). 2014. Categorization of taxa of the flora of Cuba - 2014. *Bissea* 8 (special issue 1): 5-307.
González-Oliva, L., L. R. González-Torres, A. Palmarola and D. Barrios (Eds.). 2015. Categorization of taxa of the flora of Cuba - 2015. *Bissea* 9 (special issue 4): 3-707.
González-Robledo, A., L. Robledo and A. Enríquez. 2010. Flora and vegetation of "Lomas de Galindo" Canasí, Havana. *Journal of the National Botanical Garden* 30-31: 39-50.
González-Torres, L. R. and R. Berazaín. 2004. The serpentine vegetation of lomas de La Coca, Havana City. *Journal of the National Botanical Garden* 25-26: 79-86.
González-Torres, L. R., A. Palmarola, E. R. Bécquer, R. Berazaín, D. Barrios, and J. L. Gómez- Hechavarría. 2013. Top 50: Cuba's 50 most endangered plants. *Bissea* 7 (special issue 1): 1-107.
González-Torres, L. R, A. Palmarola, D. Barrios, L. González-Oliva, E. Testé,
E. R. Bécquer, M. A. Castañeira-Colomé, J. L. Gómez-Hechavarría,
J. A. García-Beltrán, D. Rodríguez-Cala, R. Berazaín, L. Regalado and L. Granado.2016. Conservation status of the flora of Cuba. Bissea10 (NE1): 1-23.

González-Oliva, Ferro, Rodríguez-Cala and Berazaín. 2017. Methods for the inventory of populations. Chapter 8. Plant Inventory Methods. In: *Biological Diversity of Cuba. Inventory methods, monitoring and biological collections (*C. A. Mancina and D. D. Cruz Flores, Eds.). Editorial AMA, Havana, 502 pp.

Granado, L. 2015. *Population structure of* Magnolia cubensis subsp. acunae. 2015. Bachelor's thesis. Faculty of Biology, University of Havana.

Granado, L., R. Núñez, D. Martínez, S. Delfín, B. Falcón, V. Pérez, and L. R. González-Torres. 2016. Population structure of *Tabebuia lepidophylla* (Bignoniaceae) in the pine forest on quartzite sands of Los Pretiles Ecological Reserve, Pinar del Río, Cuba. *Journal of the National Botanical Garden* 37: 29-37.

Gras, M. J., J. Raventós, A. Bonet and D. A. Ramírez. 2002. Multiscalar analysis of the distribution patterns of the Alicante endemism *Vella lucentina* MB Crespo (Brassicaceae) and implications on its conservation. *Geographicalia* 42: 93-112.

Guzmán, J. M. and L. Menéndez. 2013. *Protocol for monitoring the mangrove ecosystem*. National Center for Protected Areas, Havana. 29 pp.

Hegazy, A. K. and N. M. Eesa. 1991. On Ecology, insect seed-predation and conservation of rare and endemic plant species: *Ebenusarmitagei* (Leguminosae). *Conservation Biology* 5(3): 317-324.

Henderson, A., G. Galeano & R. Bernal. 1995. Field Guide Palms Americas 1-
352. Princeton University Press, Princeton, New Jersey.

Hernández, M. and D.D. Cruz. 2016. Natural vegetation cover in Cuban National Parks: multitemporal analysis and futu127 Chapter.
8. Methods of Plant Inventory ra of bioclimatic conditions.
Journal of the National Botanical Garden 37: 93-102.

Judd, W. S. S.; C. S. Campbell, E. A. Kellogg, P. F. Stevens, M. J. Donoghue (2007). "Arecaceae. *Plant Systematics: A Phylogenetic Approach, 3rd ed*. Sunderland, MA: Sinauer Associates. pp. 278-280. ISBN 978- 0-87893-407-2.

Krebs, C. J. 1999. Ecological Methodology. 2nd ed.
Benjamin/Cummings. León, Br. and Alain , Br. 1951 . Dicotyledons:
Casuarinaceae to Meliaceae. *Occasional Contributions of the Museum*

of Natural History Colegio "De La Salle" 10: 1-424.
Leon, Br. 1946. Flora of Cuba 1. Gymnosperms to Monocotyledons. *Occasional Contributions of the Museo de Historia Natural Colegio "De La Salle"* 8: 1-405.
López, E., E. Bicerra and E. Diaz. 2006. Ecological profile of four stands of camucamu tree *Myrciaria floribunda* (H. West. ex Willd) O. Berg. in Ucayali. *Ecología Aplicada* 5: 45-52.
Martínez, E. 2014. Description of new phytocenoses in the ophiolitic plain of Camagüey province, Cuba. *Journal of the National Botanical Garden* 34-35: 19-28.
Martínez, E. and O. J. Reyes. 2015. Characterization of the vegetation of the San Felipe plateau in Camagüey, Cuba, for conservation purposes.
Journal of the National Botanical Garden 36: 19-30.
Martínez, E., Z. Acosta, D. Godínez and J. M. Plasencia. 2009-2010. New phytocenoses in the ophiolitic plain of Camagüey province, Cuba. *Journal of the National Botanical Garden* 30- 31: 141-152.
Matos, J. and A. Torres Bilbao. 2000. Early successional stages of Cuabal in the Santa Clara serpentines. Journal of the National Botanical Garden 21(2): 167-184.
Matteucci, S. D. and A. Colma. 2002. *Methodology for the study of vegetation*. CONICET, Argentina, 159 pp.
Menéndez, L., R. Capote, J. M. Guzmán, L. F. Rodríguez and A. V. González. 2006. Mangrove ecosystem health in the Sabana-Camaguey Archipelago: Patterns and trends at the landscape scale. Menéndez, L. and J.
M. Guzmán (Eds.) Ecosistema de manglar en el archipiélago cubano. UNESCO, Havana.
Menges, E.S. 1990. Population viability analysis for an endangered plant.
Conservation Biology 4(1): 52-62.
Noss, R.F. 1990. Indicators for monitoring biodiversity: a hierarchical approach.
Conservation Biology 4 (4): 355-364.
Oviedo, R. and L. González-Oliva. 2015. National list of invasive and potentially invasive plant species in the Republic of Cuba- 2015. *Bissea* 9 (special issue 2): 5-91.

Oviedo, R., M. Fernández and M. A. Vales. 1988. Floristic study of Cayo Alfiler. Toscano Farm. Pinar del Río. *Journal of the National Botanical Garden* 9(3): 75-84.

Oviedo, R., P. Herrera, M. Caluff, L. Regalado, I. Ventosa, J. M. Plasencia, I. Baró, P. A. González, J. Pérez, L. Hechavarría, L. González- Oliva, L. Catasús, J. Padrón, S. I. Suárez, R. Echevarría, I. M. Fuentes, R. Rosa,

P. O.

Rodríguez, W. Bonet, M. Villate, N. Sánchez, G. Begué, R. Villaverde, T. Chateloin, J. Matos, R. Gómez, C. Acevedo, J. Lóriga, M. Romero, I. Mesa, A. Vale, A. T. Leiva, J. A. Hernández, N. E. Gómez, B. L. Toscano,

M. T. González, A. Menéndez, M. I. Chávez and M. Torres. 2012. National list of invasive and potentially invasive plant species in the Republic of Mexico.

Cuba - 2011. *Bissea* 6 (NE 1): 22-96.

Pérez, E. and M. Armas. 1987. Effect of disturbance by mowing on savanna herbaceous vegetation. *Journal of the National Botanical Garden* 8 (1): 53-67.

Primack, R., R. Rozzi, P. Feinsinger, R. Dirzo and F. Massardo. 2001. *Fundamentals of biological conservation. Latin American perspectives.* Fondo de Cultura Económica. Mexico, DF. 797 pp.

Quero, H. J. 1994. Palmae. Fl. Veracruz 81: 1-118.

Quero, H. J. & R. W. Read. 1986. A revision of the palm genus Gaussia. Syst.

Bot. 11(1): 145-154.

Rodríguez-Cala, D. and L. González-Oliva. 2015. Invasion and impact of *Tithonia diversifolia* (Asteraceae) in the Protected Natural Landscape Topes de Collantes, Cuba. *Journal of the National Botanical Garden* 36: 151-162.

Rodríguez-Cala, D., R. Valdez, A. Dulón, M. Osés, R. Pérez, Z. Esquivel, B. Falcón, V. Pérez, and L. González-Oliva. 2017. Conservation status of *Erigeron bellidiastroides* (Asteraceae) at the new locality Los Pretiles, Pinar del Río, Cuba. *Journal of the National Botanical Garden* 38: 43-50.

Romero-Jiménez, M., L. Más, R. Oviedo, J. A. Pegudo, A. Arias and L. Morales. 2015. Status of *Scaevola sericea* (Goodeniaceae) in the

northeastern cay of Villa Clara, Cuba. *Journal of the National Botanical Garden* 36: 181-187.
Samek, V. 1973. Phytogeographic regions of Cuba. *Cuban Academy of Sciences. Serie. Forestal* 15:1-60.
Sánchez, J. A., B. Muñoz and L. Montejo. 2009a. Tree seed traits in a tropical evergreen forest of Sierra del Rosario, Cuba. *Pastures and Forages* 32(2): 141-164.
Sánchez, J. A., B. Muñoz, L. Montejo and R. Herrera. 2009b. Ecological grouping of tropical trees in an evergreen forest of the Sierra del Rosario, Cuba. *Acta Botánica Cubana* 204:14-23.
Silva, J. F.; J. Raventos, H. Caswell and M. C. Trevisan. 1991. Population responses to fire in a tropical savanna grass, *Andropogon semiber bis*: a matrix model approach. *Journal of Ecology* 79: 345-356.
Silva, J. F. and J. Raventos. 1999. Effects of end of dry season shoot renoval on growth of three savanna grasses with different phenologies. *Biotropica* 31 (3): 430-438.

Simpson, Michael G. (2005). "Arecaceae. *Plant Systematics*. Elsevier Inc. pp. 185-188. ISBN 0-12-644460-9 ISBN 978-0-12-644460-5.
Stohlgren, T. J., M. B. Falkner, and L. D. Shell. 1995. How to establish a modified- whittaker nested vegetation sampling method. *Vegetatio* 117: 113-121.
Sutherland, W. J. 1995. *Census Techniques*. Blackwell Science, Oxford. Sutherland, W. J. 2006. *Ecological Census Techniques: a handbook*.
Cambridge University Press. New York, 432 pp.
Testé E., L. González-Oliva and A. Márquez. 2015. Current and potential invasion of the toxic tree *Rhus succedanea* (*Anacardiaceae*) in the Protected Natural Landscape Topes de Collantes, Cuba. *Journal of the National Botanical Garden* 36: 173-180.
Turner, W., S. Spector, N. Gardiner, and M. Flodely. 2003. Remote sensing for biodiversity science and conservation. *Tree* 18: 306-314.
IUCN (International Union for Conservation of Nature). 2001. IUCN Red List Categories and Criteria: version 3.1. IUCN Species Survival Commission. IUCN, Gland, Switzerland and Cambridge, UK. ii + 33 pp.
Urquiola, A., L. González-Oliva, R. Novo and Z. Acosta. 2010. *Red book of the vascular flora of Pinar del Río province*. Publications

University of Alicante, Alicante, 457 pp.
Zippel, E., T. Wilhalm and C. Thiel-Egenter. 2010. Methods for sampling higher plants. Pp. 346- 376. In: *Manual on field recording techniques and protocols for all taxa biodiversity inventories and monitoring* (Eymann, J.,
J. Degreef, C. Häuser, J.C. Monje, Y. Samyn and D. Vanden Spiegel, eds). ABC Taxa, vol 8, 653 pp.

Printed by Books on Demand GmbH, Norderstedt / Germany